DATE DUE			

POWERFUL
Waves

A copy of the *Great Wave Off the Coast of Kanagawa* by the famous Japanese artist Hokusai

POWERFUL Waves

by D. M. Souza

Carolrhoda Books, Inc./Minneapolis

*For everyone who enjoys
watching ocean waves*

With special thanks to editor Marybeth Lorbiecki for her many insightful questions and suggestions, and to Ms. Patricia Lockridge, coauthor of *U. S. Tsunamis: 1690-1988*, for her assistance with this book.

METRIC CONVERSION CHART		
To find measurements that are almost equal		
WHEN YOU KNOW:	MULTIPLY BY:	TO FIND:
feet	30.48	centimeters
yards	0.91	meters
miles	1.61	kilometers

Text copyright © 1992 by D. M. Souza
Illustrations copyright © 1992 by Carolrhoda Books, Inc.

Library of Congress Cataloging-in-Publication Data
Souza, D. M. (Dorothy M.)
   Powerful waves / by D. M. Souza.
      p.   cm. — (Nature in action)
   Summary: Facts about ordinary waves precede information about the causes of the huge waves known as tsunamis and the destruction they bring.
   ISBN 0-87614-661-2
   1. Tsunamis—Juvenile literature. [1. Ocean waves. 2. Tsunamis.]
I. Title. II. Series: Nature in action (Minneapolis, Minn.)
GC221.2.S68   1992
551.47′024—dc20   91-885
                                                                      CIP
Manufactured in the United States of America   AC

1 2 3 4 5 6 7 8 9 10 01 00 99 98 97 96 95 94 93 92

Contents

Acknowledgments:
Front cover photograph by Jeff Greenberg; Back cover photograph courtesy of The Mansell Collection; A. Quinton White/Visuals Unlimited, p. 1; Independent Picture Service, pp. 2, 9; Jeff Greenberg, pp. 4, 14 (top & bottom), 37, 42 (right); U.S. Navy, p. 5; Maria Veghte/Visuals Unlimited, p. 6 (left); Jerg Kroener, p. 6 (right); Bruce Berg/Visuals Unlimited, pp. 7 (inset), 11; Ken Gosner/Visuals Unlimited, p. 7; U.S. Army Corps of Engineers, pp. 8, 22 (bottom); Dave B. Fleetham/Visuals Unlimited, p. 13; Peter K. Ziminski/Visuals Unlimited, p. 15 (top & bottom); John Gerlach/Visuals Unlimited, p. 16; UPI/Bettmann Newsphotos, p. 17; The Mansell Collection, pp. 18-19, 31; NOAA/EDIS, p. 20; U.S. Coast Guard, p. 21 (left & right); NOAA, pp. 22 (top), 35; Kjell B. Sandved/Visuals Unlimited, pp. 23 (left), 26; Charles S. Houston/Visuals Unlimited, p. 23 (right); International Tsunami Information Center, p. 27 (left & right); Greg Vaughn/Hawaii Visitors Bureau, p. 28; Mrs. Harry S. Simms, Sr., p. 32 (top & bottom); U.S. Department of the Interior, p. 33 (top & bottom); Daniel Gotshall/Visuals Unlimited, p. 34 (inset); SATOUR, pp. 34, 48; Takaaki Uda, Public Works Research Institute, Japan, pp. 36, 38 (right); Les Christman/Visuals Unlimited, p. 38 (left); Pacific Operations Group/NOAA, pp. 40, 42 (left); Alaska Tsunami Warning Center, p. 41 (left & right); American Lutheran Church, pp. 44-45; Minneapolis Public Library and Information Center, p. 46; U.S. Geological Survey, p. 47. Illustrations on pp. 10-11, 12, 14, 25, 26, 29 by Bryan Liedahl. Maps on pp. 24, 39 by Laura Westlund. Map on p. 39 based on information provided by the International Tsunami Information Center. The quotation about the 1960 tsunami that hit Hilo, Hawaii (p. 32), first appeared in the National Oceanic and Atmospheric Administration (NOAA) magazine of January 1974 (Vol. 4, No. 1).

If you've ever walked along an ocean beach, you know how restless the sea is. Waves break against the shore, leave treasures from the deep, then slip back into the sea again.

Sometimes the waves are gentle. Other times they foam and churn, tossing ocean ships about like toy boats.

One kind of wave, however, is more powerful than all others. It can tear apart homes, steel bridges, and concrete buildings. It can uproot trees, twist power lines, and sweep city blocks clean of everything. This wave is a tsunami (soo-NAH-mee).

Tsunami is a Japanese word meaning "large waves in the harbor." The word can mean a single wave or a series of waves. The Japanese people know the force of these waves well. They have seen thousands of people, their homes, and their belongings washed away by tsunamis.

Some people call these mighty waves "tidal waves." But as you will see, tsunamis have little to do with the tides.

A Japanese artist illustrated a tsunami in the tradition of Hokusai

9

What Are Waves?

In the sea, waves are up-and-down movements of water. Small waves, called ripples, only lift the water inches above the surface. Other waves, like storm waves, may raise water over the decks of ships.

The highest point of a wave is its crest. Between the crests of two waves is a low valley called the trough. The height of a wave is the distance from the trough up to the crest.

Scientists can tell how long a wave is by measuring how far one crest is from the next. Inches may separate the crests of ripples. The length of several football fields may separate the crests of larger waves.

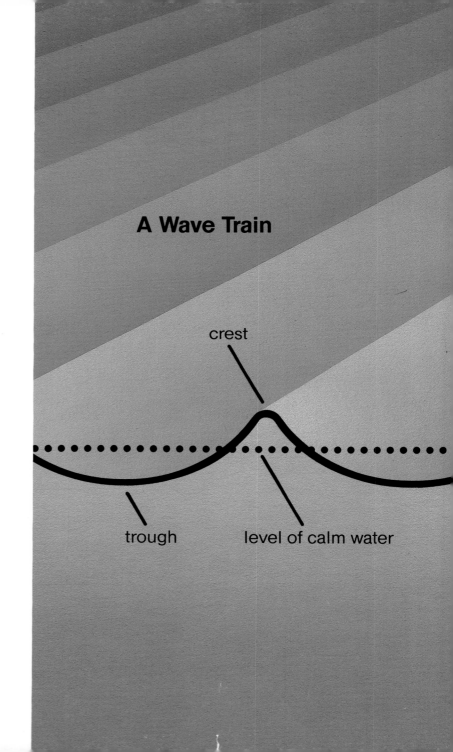

A Wave Train

crest

trough

level of calm water

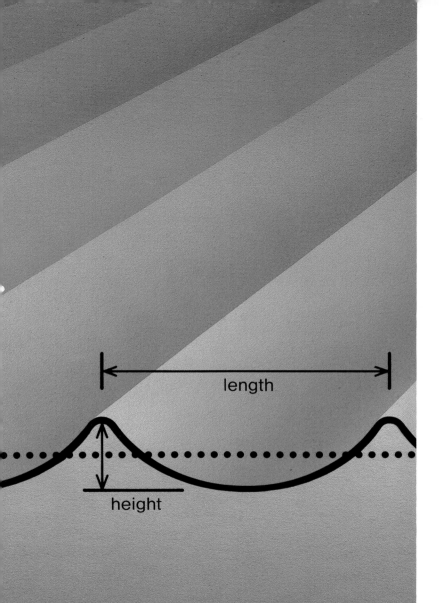

length

height

Waves usually follow one another, forming a train of waves. In the ocean, many trains are moving in different directions at the same time. Some crisscross one another. Their waves may run together and grow bigger. Or the trains may break apart, making the waves disappear.

The time it takes two crests in a train to pass the same point is known as a wave period. Scientists use wave periods to figure how fast waves are moving. Most waves at sea have periods of 2 to 20 seconds. This means the waves are traveling from 7 to 70 miles per hour. The average speed of an ocean wave is about 35 miles per hour.

In Deep Water

When you watch waves far out at sea, you may think the water is moving along with the waves. But it isn't. Your eyes are playing tricks on you.

Try tying a rope to a doorknob and snapping the rope several times. Notice how it lifts and falls in wavelike motions. The waves move along the rope, but the rope does not move forward. This is what happens with water waves.

Imagine water lying under the snapping rope. With each up-and-down movement of the rope, the water also moves up and down. It rises to a crest, turns a somersault into a trough, and then returns to about the same spot it was when it began.

Beneath the first large somersault, the water turns in smaller somersaults—one beneath the other. Each turn is deeper and smaller.

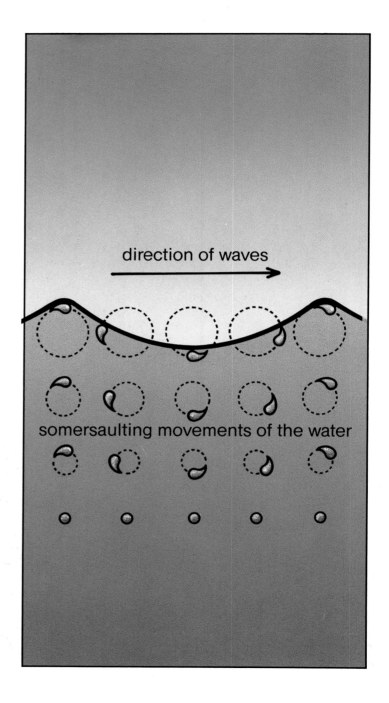

direction of waves

somersaulting movements of the water

This photograph shows what waves look like under the surface of the water. One can see the puffy pattern made by the water moving in many somersaults.

Near Shore

As a wave train gets closer to shore, things change. The ocean gets shallower. Water below the waves has less and less room to turn in somersaults. Some somersaults strike the ocean bottom and drag. Then the water simply moves back and forth in a straight line.

The water at the top, though, continues to somersault. It turns over faster than the water below. The crests begin to crowd together. They build higher and higher until they become top-heavy. Finally they tumble over. A foaming mass of water rushes toward shore, races onto the beach, then slips back into the sea.

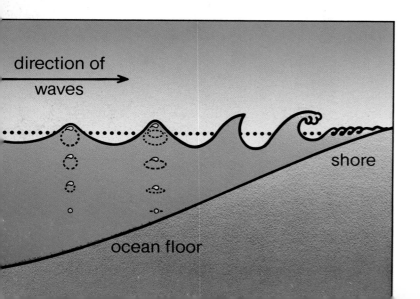

direction of waves

shore

ocean floor

How Do Most Waves Form?

High tide

Waves are usually formed by the movements of the tides and the winds.

Tides are the work of the moon and the sun. The moon and the sun pull on the earth like a magnet. This pulling raises the level of the earth's water in some places and lowers it in others. Twice each day, water comes farther up onto shore than normal. This is called high tide. At low tide, the water is pulled away from the shore, and moves farther back into the sea.

If a storm arrives during a high tide, waves on shore will be stronger and higher. High tides can bring dangerous floods sweeping inland.

Low tide

Lakes are pulled by the sun and moon too. But since lakes are smaller than oceans, tides are smaller and harder to notice. The moon shines here over Mono Lake in California.

Most waves in oceans and lakes are caused by the wind. A slight breeze blows across calm water, and ripples appear. If the wind picks up speed and strength, the ripples become bigger waves.

Winds blowing at 30 miles per hour can lift the sides of waves to about 15 feet.

Strong, wind-driven waves rise to an average height of 40 to 50 feet. Hurricane winds can whip up even stronger superwaves that may sweep away buildings, animals, cars, trees, and power lines.

Some people think these large storm waves are tsunamis. They are not.

What Are Tsunamis?

Tsunamis are caused by earthquakes, volcanoes, landslides, or avalanches that rock the ocean. After these violent movements, the sea may pitch and swirl beneath the water's surface. Then long waves move outward from the area in all directions. These waves are faster and more powerful than any made by the tides or the wind.

Because earthquakes cause most tsunamis, the waves are often called seismic sea waves. *Seism* [SY-zum] means earthquake.

An earthquake shook Lisbon, Portugal, in 1755, causing a tsunami. This was the first tsunami to be studied scientifically, and it was one of the few tsunamis to have struck in the Atlantic Ocean.

Unlike wind-driven waves, tsunamis are not high as they travel across the open sea. They may be only a few feet from trough to crest. Most of the time, sailors at sea do not even notice them.

In 1946, a freighter was anchored a short distance off the coast of Hilo, Hawaii. Glancing toward shore, the crew suddenly saw huge waves crashing onto the city. Trees fell and buildings collapsed. Piers, boats, and 12-ton rocks were carried on land by the charging waters. About 160 people lost their lives. It was the worst natural disaster in Hawaii's history. Yet the men had felt nothing when the low waves passed beneath their ship.

The 1946 tsunami that struck Hilo, Hawaii

In 1946, a strong quake struck off the coast of Alaska. On nearby Unimak Island stood the towering Scotch Cap Lighthouse, built to withstand the pounding seas around it. The lighthouse's foundation was 32 feet above sea level.

Twenty minutes after the quake, a 100-foot-high wave smashed into the lighthouse. All that remained was twisted steel and shattered concrete. This tsunami was the same one that later hit Hilo, Hawaii.

Tsunamis are very long waves. Their crests may be 200 miles apart. This is about as far as Boston is from New York.

The water in these long waves churns in somersaults deep in the ocean. The deeper the water, the faster tsunamis go. In water 500 feet deep, they move about 87 miles per hour. In water 18,000 feet deep, their speed may reach 519 miles per hour. Powerful tsunamis traveling through the deepest part of an ocean have reached speeds of 600 miles per hour.

When a tsunami nears shore, the shallowness of the water acts like a brake. Suddenly the speed of a wave may drop to 40 miles per hour. The water beneath the wave piles up. Crests rise higher and higher. In seconds, a 2-foot-high wave is 30 feet high. It arches and crashes onto shore with a terrible roar.

And this may not be the end. There may be more waves in this tsunami—waves that jam together and build up even higher than before.

Since the force behind a tsunami—an earthquake, for example—is much stronger than any wind, these mighty waves usually travel longer distances than other waves.

In 1960, several violent earthquakes shook the ocean floor near Chile. The land along the coast rose and fell, and the sea pitched wildly. Huge waves tore apart boats and then flooded the land, carrying entire villages out to sea. Between 1,000 and 1,500 people drowned.

Meanwhile, other waves in the tsunami raced north across the Pacific Ocean. Some of these waves smashed boats and harbors near the coast of southern California.

Fifteen hours after hitting California, the tsunami reached Hawaii and left cities in ruins. In twenty-two hours, the tsunami arrived at Japan, where it destroyed 5,000 homes and claimed the lives of almost 200 people. The tsunami also left its mark on places as far away from Chile as New Guinea, New Zealand, the Philippines, and Pitcairn Island.

The 1960 tsunami raced across the Pacific Ocean from Chile, causing incredible damage. This is the coastal area of Isla Chiloe, Chile, after the event.

The Waiakea area of Hilo, Hawaii, after the 1960 tsunami

22

How Do Tsunamis Form?

The oceans of the world are extremely deep. Along their bottoms are mountains higher than any found on earth. Some underwater mountain peaks are so high they rise above the surface of the water as islands. Hawaii is a range of mountain peaks in the Pacific Ocean.

Huge valleys and long, narrow canyons, or trenches, also line the bottom of the oceans. The deepest-known canyon is the Mariana Trench in the Pacific Ocean. If Mount Everest (the tallest mountain on dry land) were dropped into this trench, you would not even see a trace of it.

From time to time, mountains in the oceans shake and crack. Trenches collapse, and volcanoes erupt. Scientists believe these things happen because the oceans and the continents rest on huge stone plates. Each plate is a massive slab of rock, thousands of miles wide. Some plates are up to 80 miles thick. More than 330 Empire State Buildings would have to be stacked on top of one another to equal the thickness of these plates.

The Hawaiian Islands are actually the peaks of an enormous underwater mountain range.

Mount Everest

Plates Under the Pacific Ocean

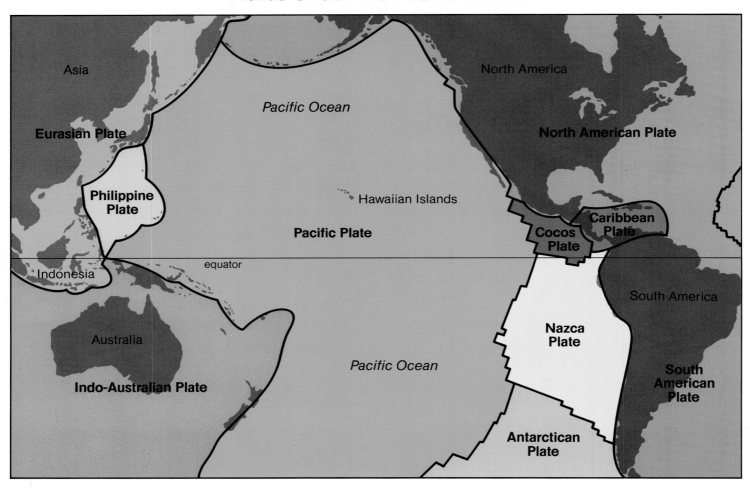

About seven large plates and many smaller ones cover the earth. The plates fit together like the pieces of a giant jigsaw puzzle. They float like rafts on churning, molten rock called magma. Each year, the plates move a few inches in different directions. The Pacific Plate moves the most. This is why so many earthquakes, volcanic eruptions, and tsunamis take place in the Pacific.

Earthquakes

You know what happens when you jiggle your hand around under water. Waves appear on the water's top.

Imagine what happens when gigantic undersea plates move in an earthquake. The ocean floor rises and buckles. Or gaps open up, swallowing rocks and huge chunks of earth. The ocean water churns, and powerful waves form. The waves belong to a tsunami.

In 1896, a strong earthquake shook the bottom of the sea 93 miles off the coast of Japan. The quake was hardly felt on land.

But 20 minutes after it struck, the ocean began pulling back from the beaches.

Most of the people in the villages along the coast were celebrating a festival, so they didn't notice anything unusual on the beaches. But low waves were racing toward them as fast as jet planes.

The closer the waves came to the shore, the larger they grew. One wave rose higher than a 10-story building. It crashed onto the beach and washed over the villages. Twenty-seven thousand people drowned, and 10,000 homes were destroyed by this tsunami.

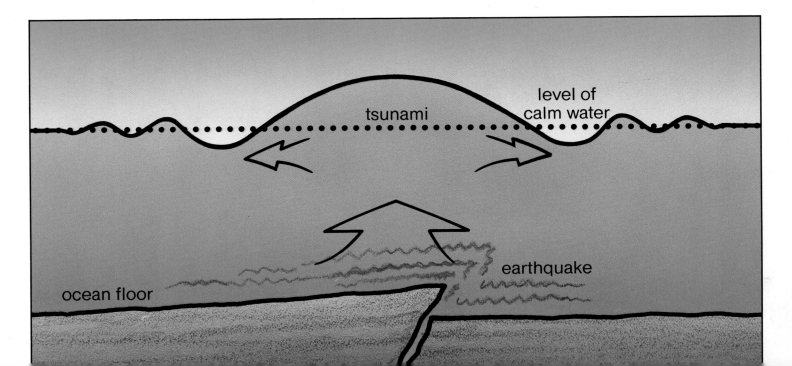

Landslides and Avalanches

If you drop a stone into a glassy pond, waves form. The water ripples away from the spot in a series of widening circles.

Sometimes tons of rock or mud fall into the ocean. Other times, great chunks of snow or ice slide off mountains and crash into the water. Then tsunamis that are a million times stronger and faster than ripples move outward in enormous circles.

From the sky, one can see waves moving out in widening circles from a place where the ocean has been disturbed.

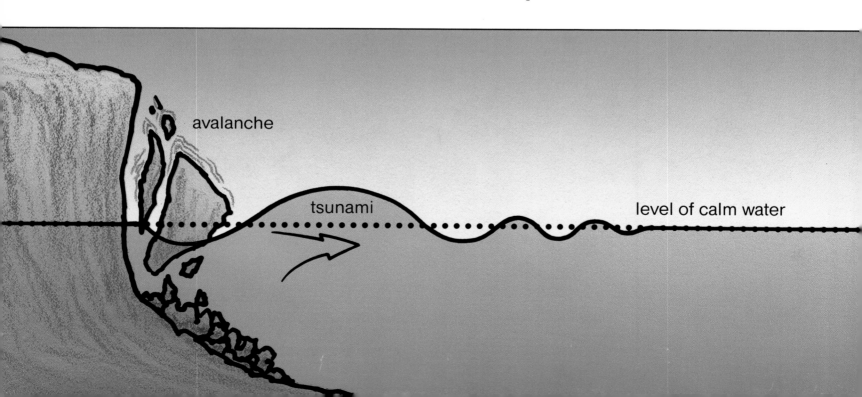

avalanche

tsunami

level of calm water

The Alaskan tsunami of 1964 surged over towns on the coast, overturning trucks, boats, buildings, and trees.

In 1964, one of the strongest quakes ever recorded shook Prince William Sound, Alaska. Its power was 12,000 times greater than the atom bomb dropped on Hiroshima. Sixty-foot-high buildings fell as if they were built of popsicle sticks. The editor of a newspaper in Kodiak wrote that when he tried walking, it was like "marching across a field of Jell-O."

Minutes after the quake, tons of earth along the Alaskan coast fell into the sea. Water in the Gulf of Alaska began churning wildly, and tsunamis raced in all directions. Waves 90 feet high washed over towns on the coast, while other waves moved out to sea. The tsunamis caused major damage along the West Coast of North America as well as in Hawaii.

Volcanoes

Volcanoes are openings in the earth through which magma escapes. The openings often lie near places where two plates meet. Sometimes when the plates move, one slips over the other. The bottom plate plunges into the hot, molten rock deep inside the earth. The magma explodes up, and a volcano erupts.

At other times, two plates shift suddenly away from each other. The molten rock rises up between them and shoots through a volcano into the air.

As magma blasts out of the earth, part of the volcano may break away and fall into the sea. Tons of ash may drop from the sky into the water. When these things happen, tsunamis form, and waves race outward in every direction.

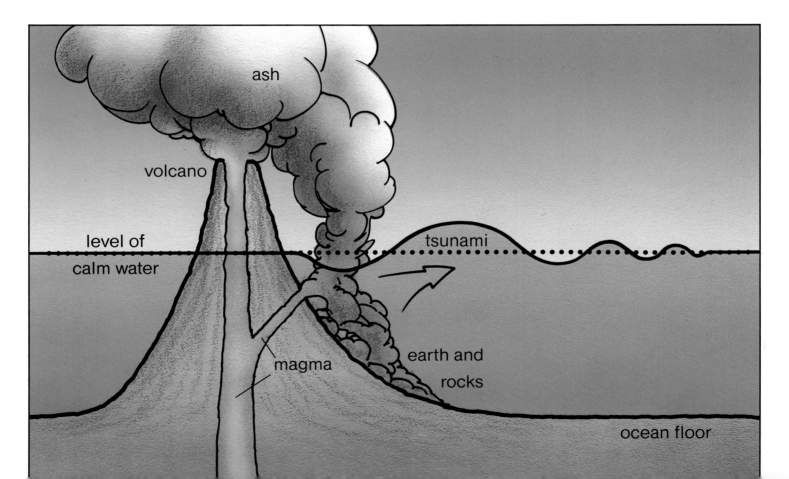

Volcanoes exist in places around the world. But the greatest number of eruptions take place in a circle around the Pacific Ocean. The area is known as the Ring of Fire. This is where many tsunamis start.

One of the most powerful tsunamis ever recorded was caused by a volcano on the eastern rim of the Ring of Fire. Early one August morning in 1883, the island of Krakatoa, in Indonesia, was rocked by the movement of an undersea plate. The island's volcano exploded with a noise that could be heard thousands of miles away. Even people in the middle of Australia heard the explosion.

Dust and ash blew miles into the sky and began circling the earth. It was so dark on nearby islands that people had to light their lamps at noon.

The eruption of Krakatoa triggered a tsunami. The waves moved at breathtaking speed onto the shores of Java and Sumatra. More than 36,000 people drowned. Thousands of homes and villages were destroyed. Even boats in harbors hundreds of miles away were sunk by the waves. The tsunami had so much power that the waves circled the globe for days.

What Do Tsunamis Look Like?

Before the Chilean tsunami hit Hilo, Hawaii, three scientists watched the ocean from a safe lookout. This is what they saw and heard:

A dull rumble like a distant train came from the darkness. Our eyes searched for the source of the noise. In the dim light, we saw a pale wall of tumbling water. It grew higher as it moved steadily toward the heart of the city. A 20-foot-high wall of water churned past our lookout. Seconds later, brilliant blue-white flashes lit up the darkness as the wave washed over the town with crushing force, snapping off power poles, grinding buildings together, and flooding the city.

Height

Each wave in a tsunami has a different height and strength when it arrives on shore. And each wave acts differently at different stretches of coast.

Sometimes a tsunami arrives as a tall wall of water. The waves may rise 100 feet or more. The sound is deafening. Crests curl over into each other. Then an enormous foaming monster crashes and races over the land.

At other times, a tsunami hits shore as a charging flood of water. It pushes forward like a bulldozer, smashing everything in its path.

Before tsunamis arrive on shore, they may slam into undersea mountains and valleys, steep cliffs, islands, and coral reefs. These collisions can make the waves stronger or weaker. A tsunami may be 50 feet high on one shore. A few miles away, the same wave may be only 5 feet high.

Tsunamis are changed by the tides too. A small tsunami always grows larger if it comes ashore with a high tide.

Tsunamis can also be carried away from coastlines by powerful ocean currents. The waves then move around the sea for days until they lose their power.

Number and Timing

As many as eight waves may be in one tsunami. Usually the first wave is not the highest or most powerful. The greatest height or strength may come with the second or third wave, or maybe even with the seventh or eighth.

The waves in a tsunami do not always follow each other quickly, one after the other. Ten to sixty minutes may pass between the arrival of waves. No one can predict how many waves will crash onto a shore, or how far apart each will be.

After the enormous earthquake that rocked Alaska in 1964, a tsunami raced south toward the West Coast of the United States. Two waves flooded streets in Crescent City, California, and then retreated. Several people, thinking it was safe, returned to the scene to see what had happened. Unfortunately, they were taken by surprise when an even larger wave came in and carried them out to sea.

These photographs show the parking lot of a Japanese aquarium before, during, and after a tsunami flooded it in 1983. Notice how far the water pulled back from the beach before the tsunami rushed in.

Signs of a Tsunami

Before a tsunami reaches land, it may give its own warning. Strange sights or sounds on the beach often signal the approach of a tsunami.

Before the arrival of the giant waves in Hawaii in 1946, eyewitnesses heard hissing noises as foaming water swept over the beach. Then the water pulled back into the sea much farther than it had ever done during a low tide. Hundreds of fish were left flapping on the sand. Minutes later, like a boomerang, a great flood of water returned, hitting the shore. It buried highways and homes. Many people drowned.

The ocean frequently pulls back from the beach before the arrival of a tsunami. The water may also thunder, crash, and foam. When these things happen, danger is racing across the sea at full speed.

Where Do Tsunamis Happen and How Often?

Tsunamis have formed in the Atlantic Ocean, Indian Ocean, the Caribbean and Mediterranean Seas, and other bodies of water. However, the greatest number of tsunamis have started in the Pacific Ocean.

Fortunately, the powerful waves do not happen after every volcanic eruption, earthquake, landslide, or avalanche that happens near a coast. In June 1991, Mount Unzen in Japan erupted violently. It sent up clouds of steam and ash 20 miles into the sky. But no tsunami appeared in the Pacific.

About five tsunamis are reported yearly around the world. Most of these are small and do little damage. Destructive tsunamis seem to strike somewhere in the world about once every 10 years.

Recently there have been more than 1 every 10 years. Hawaii was hit by a destructive tsunami in 1975, and parts of the Philippines were washed away in 1976.

In 1983, after a major quake in the Pacific Ocean, 50-foot waves hit northern Japan.

One hundred people were killed, more than 1,500 homes were destroyed, and hundreds of boats were sunk or damaged.

Another quake, in Mexico in 1985, sent waves crashing against that country's Pacific coast. No one was killed, but the damage to property was heavy. Oyster beds and a farm for endangered turtles were among the things destroyed.

In 1988, a tsunami washed away 13 villages in the Solomon Islands. No one can be certain when or where the next tsunami will strike.

A tsunami that struck Japan in 1983 left many coastal areas in ruins.

Tsunamis Recorded from 1690 to 1990

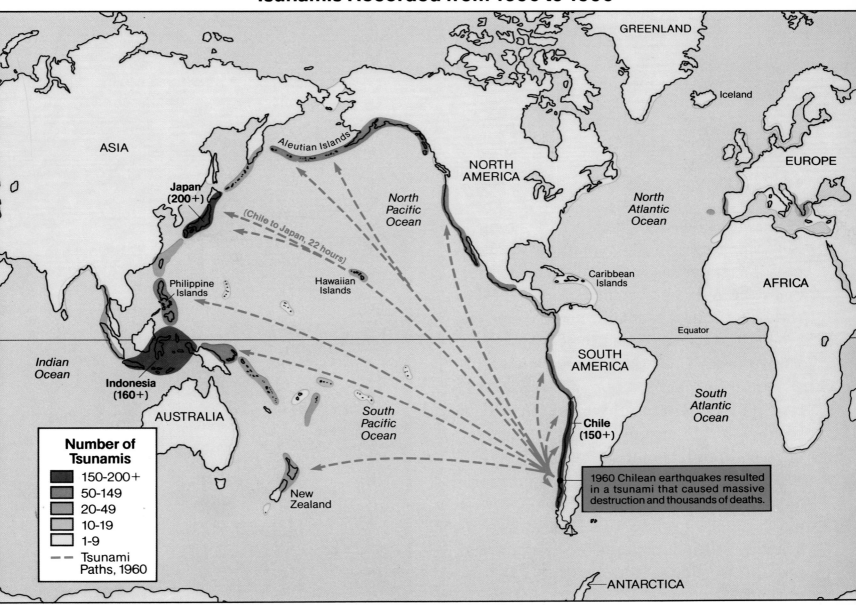

GREENLAND

Iceland

ASIA

Aleutian Islands

Japan
(200+)

NORTH
AMERICA

EUROPE

North
Pacific
Ocean

North
Atlantic
Ocean

(Chile to Japan, 22 hours)

AFRICA

Philippine
Islands

Hawaiian
Islands

Caribbean
Islands

Indian
Ocean

Equator

Indonesia
(160+)

SOUTH
AMERICA

AUSTRALIA

South
Pacific
Ocean

Chile
(150+)

South
Atlantic
Ocean

**Number of
Tsunamis**

150–200+
50–149
20–49
10–19
1–9

Tsunami
Paths, 1960

New
Zealand

1960 Chilean earthquakes resulted
in a tsunami that caused massive
destruction and thousands of deaths.

ANTARCTICA

39

Warning Systems

In 1946, scientists began setting up a warning system that would protect people from tsunamis. It was called the Seismic Sea Wave Warning System (SSWWS). After the 1960 Chilean tsunami, many countries around the Pacific joined the SSWWS.

Five years later, the SSWWS grew into the International Tsunami Warning System (ITWS). Twenty-three countries now belong to this organization. Its central offices are in Honolulu, Hawaii.

Scientists from member nations work on various islands and in coastal towns and villages around the Pacific Ocean. They manage 69 earthquake observatories, 65 tide stations, and 101 information-gathering centers.

Whenever major earthquakes, volcanoes, landslides, or avalanches happen, experts pinpoint the center and strength of the earth movements. They use instruments called seismographs. If the activity is near the ocean, a Tsunami Watch is sent by satellite to the people along the ocean's coasts. A Tsunami Watch is a message telling people that powerful waves may be forming. The message takes three minutes to arrive.

News broadcasters on TV and radio interrupt regular programs to announce the Tsunami Watch. In cities near the movement's center, sirens also blare.

Meanwhile, scientists at nearby tide stations study the waves. They use instruments called tide gauges to measure the waves' heights and speeds. Low, fast-moving waves mean a tsunami has formed. If any of the tide gauges stop working, the experts begin to worry. They figure the gauges have been washed away by a strong tsunami.

Using special charts, the scientists quickly decide how long the tsunami will take to reach different shores. Remember, the speed of a tsunami depends in part on how deep the water is. The charts show the depths of the ocean floor.

A Tsunami Warning is then sent out. The warning means that a tsunami is definitely on its way. Sirens scream out the new message, while TV and radio announcers advise people what to do. The police and emergency workers help everyone move away from low coastlines.

When tide stations finally report that the powerful waves have disappeared, an "All Clear" signal is delivered. People then know it is safe to return to their homes.

Sometimes a quake strikes near a coastal area, and a tsunami hits shore while international scientists are still gathering information. Several times, tsunamis have struck Alaska and Japan within minutes of quakes.

To try to prevent deaths from tsunamis close at hand, local warning systems have also been set up in many areas around the Ring of Fire. Each local system has its own earthquake observatories and tide stations. When a tsunami forms close to shore, a Tsunami Watch is not sent. A warning alert sounds right away, and people know they have only a few minutes to reach safety.

Safety Tips

If you ever visit a town or city on the edge of an ocean, you should know about tsunamis. Violent activity in the land under the sea can trigger these terrifying waves at any time.

Find out ahead of time what the local tsunami signals are and what they mean. Each city or town has its own way of letting people know of the danger.

If a Tsunami Watch is given, turn on a radio or TV, and listen closely for further news. Stay away from the water. If you are in a boat, head out into deep open water. Get at least two miles from the closest stretch of shore.

When a Tsunami Warning is sounded, leave the area as quickly as you can. News announcers will tell you how much time you have, where to go, and what routes to take. Follow directions. Remain as calm as you can.

Wait for the All Clear signal before you return. It may be 24 hours before all the waves in a tsunami disappear. The ocean may appear calm, but this does not mean the danger has passed.

If an earthquake begins and you have to hang onto something to keep from falling, move away from the ocean as soon as the shaking stops. Powerful waves may wash over the land in 10 minutes or less. If the land is flat, they can rush over two miles inland in minutes. Don't wait for a warning signal. Move with lightning speed.

Once in a while, a tsunami warns you itself. You may be building sand castles or collecting shells along the beach. Suddenly, water floods the shore. Just as fast as the water has come in, it pulls back farther than you have ever seen it do before. Or maybe you hear a loud hissing and see the water churning. Move to high ground as far away from the beach as possible. Even if the first wave on shore is only a few feet high, do not be fooled. A dangerous tsunami is speeding toward you.

The power of a tsunami is like that of no other wave in the ocean. Some of the worst disasters and most spectacular events in history have been caused by tsunamis. Millions of people have lost their lives. Homes have been destroyed, and whole towns have been washed away.

As long as forces in the earth continue to stir the oceans, these fast-moving waves will race across the water. There is no telling when they will strike or where they will leave their marks of destruction.

Unusual Tsunami Events

According to some scientists, a comet or asteroid hit the earth millions of years ago. The collision was so great it set all of the earth's waters in motion. A huge tsunami is thought to have washed over the land and destroyed many living things, including the dinosaurs.

Legends from people living on several Pacific islands tell of "waves that shine in the dark." Strange lights have been spotted in the waters at night before the arrival of some tsunamis. Scientists now believe that the lights may be from the glowing bodies of tiny sea creatures called flagellates. When the ocean floor rises and falls in an undersea quake, millions of these creatures rise to the surface. They get caught in the crests of tsunamis, and their pinpoint lights make the waves glow.

Coral reefs and other undersea communities are often disturbed, smashed, or washed away by tsunamis. This can be good for these communities because it opens them up to new growth and more variety.

In 1771, a tsunami tore huge heads of coral from reefs on one of Japan's Ryukyu Islands. The coral weighed about 750 tons, more than the weight of seven blue whales. The waves carried the coral several miles inland where they can still be seen today.

In 1868, a tsunami struck the coast of Chile and Peru. A Civil War gunboat, the U.S.S. Wateree, was anchored in the harbor. The flat-bottomed ship was lifted by the waves and carried over the tops of buildings. Finally, it was dropped near the foot of a mountain. It was upright and undamaged, and only one seaman was lost.

On the morning of April 1, 1946, in Hawaii, a husband and wife were fixing breakfast in their kitchen. A tsunami suddenly arrived and swept the couple and their home several hundred feet away. When it set them down in a field of sugarcane, their breakfast was still in place.

During the same Hawaiian tsunami, a flood of water carried an electric-company worker down the main street. Buildings on either side of him collapsed, and cars were swept along beside him. At last the wave slipped back into the ocean, leaving the worker safely in a field. He had screamed so loudly during his wild ride that he could not speak for three months.

The 1964 Alaskan tsunami struck parts of Western Canada, tearing houses from their foundations. Some people clung to floating pieces of their homes to keep from drowning. When a rescue worker arrived, he saw a mattress drifting in the water, and on it was a baby, fast asleep.

A tire found in Whittier, Alaska, after the 1964 tsunami

Glossary

Crest: The highest point of a wave

Magma: Hot, molten rock under the earth's plates

Plates: Massive slabs of rock beneath the earth's land and sea

Ring of Fire: A circle of active volcanoes around the Pacific Ocean

Seismic Sea Waves: Tsunami waves caused by earthquakes

Seismograph: An instrument that records earth movements

Tidal Wave: A misleading name for a tsunami

Tide Gauge: An instrument in the water that measures wave height and period

Tides: The rise and fall of the sea due to the pull of the moon and the sun. During high tide, the water piles up along coastlines. During low tide, it moves away from coastlines.

Trough: The lowest point of a wave

Tsunami: Long, fast, powerful waves caused by earthquakes, volcanoes, landslides, or avalanches near or under the sea

Tsunami Warning: An alert telling people that a tsunami is approaching and that they should move to a safe place

Tsunami Watch: An alert letting people know that a tsunami may be forming and that they should listen for further news

Wave Height: The distance from the trough of a wave up to its crest

Wave Length: The distance from the crest of one wave to the crest of the next wave

Wave Period: The time it takes two wave crests to pass the same point or marker. This measurement is used to estimate wave speed.

Wave Train: A series of waves going in the same direction